There are many unusual stories of persons seeing the future, and some who have seen the past. Some stories seem to show that persons have actually visited the past and interacted with the people they met.

I've had many paranormal experiences myself and visions of the future which I wrote about in two of my own books. I'm also an Engineer and think this gives me a pretty unique perspective about these phenomena.

In this book you will read about many different researched cases of people visiting both the past and the future. From the evidence, this seems to be a much more common occurrence than was previously thought.

There is also a chapter on theories of how these phenomena might exist. That these might be a type of paranormal experience and/or involve existing time warps.

The world is truly much stranger than we can even imagine. I hope you enjoy these stories and they give you lots of food for thought.

I0480351

Real Time Travel Stories From a Psychic Engineer

Real Time Travel Stories From a Psychic Engineer

Real Time Travel Stories From a Psychic Engineer

Other books by Martin K. Ettington

Real Time Travel Stories From a Psychic Engineer

The Space and Aliens Six Books Bundle
A Theory of Ancient Prehistory and Giant
Aliens

The Space Colonies and Space Structures
Coloring Book
All About Asteroids

The Longevity Training Series

(A transcription of the online Multimedia Longevity Coaching Training Program)

The Personal Longevity Training Series-Book1-Long Lived Persons
The Personal Longevity Training Series-Book2-Your Soul's Purpose
The Personal Longevity Training Series-Book3-Enable Your Life Urge
The Personal Longevity Training Series-Book4-Your Spiritual Connection
The Personal Longevity Training Series-Book5-Having Love in Your Heart
The Personal Longevity Training Series-Book6-Energy Body Health
The Personal Longevity Training Series-Book7-The Science of Longevity
The Personal Longevity Training Series-Book8-Physical Body Health
The Personal Longevity Training Series-Book9-Avoiding Accidents
The Personal Longevity Training Series-Book10-Implementing These Principles

The Personal Longevity Training Series-Books One Thru Ten

These books are all available in digital and printed formats from my
website and on Amazon, Barnes & Noble, Apple ITunes, and many other sites

My Books Website is: http://mkettingtonbooks.com

Signup for our Mailing List to get the following:

1) A discount coupon for 25% discount on all books on our site

2) Occasional Notices of new books available

3) Occasional Email on other offerings of ours (Monthly)

Go to this link to sign-up:

http://personal-longevity.com/mkebooks/emailsignup/

And click this link to get the FREE 102 page Ebook titled "Secrets of Many Things"

If you have any questions about this book or other subjects please contact the Author at:

mke@mkettingtonbooks.com

Real Time Travel Stories From a Psychic Engineer

Real Time Travel Stories From a Psychic Engineer

Table of Contents

1.0 Introduction

I've always been interested in unusual subjects and wanted to learn how the world really works. I studied the paranormal in my teenage years and became an Engineer at Rensselear Polytechnic Institute in 1977 with a B.S. in Engineering Science. I also developed myself spiritually and had many paranormal and premonitional experiences.

My lifelong fascination with the unusual and unbelievable has led me to write about a series of subjects. And I have multiple books on each subject. Here are the main categories I've written about:

- The Paranormal
- Longevity and Immortality
- Ancient History & Giants
- Spirituality
- Aliens and Space

These interests have led me to write this latest book on real time travel stories and my precognitive experiences.

Since I've also had a number of visions of the future I've learned some significant facts about seeing the future. So I think I have a useful point of view to add to the review of time travel or time slip experiences.

First, I'm relating some of my most significant prophecy experiences. Then a number of cases I've researched of valid time slip experiences.

Finally, I provide my theories and explanations of how these experiences might actually work.

For many years the idea of time travel has been relegated to science fiction stories and was not considered possible by most scientists.

After reading all of these stories I now believe that time travel or "slips" do occur and we still have a lot to learn about these experiences.

Hope you enjoy the ride, since these are some really interesting stories.

2.0 My Premonitional Experiences

I've written two books about prophecy and premonitions and this gives me some insight into time travel experiences. My books are titled the following:

- Prophecy: A History and How to Guide
- Use Intuition and Prophecy To Improve Your Life-By An Adept

Here are a few of my stories. Later in this book I'll provide some theories on how time travel works.

2.1 Stories of My Major Premonitions

In this chapter I'm sharing some of my major personal experiences. These experiences provide a basis for my subjective analysis of how prophecy works.

These examples are also intended to show the reader that I do know what I'm talking about when it comes to having "sensed" the future.

a.   Visions

During the summer of 1975 I had a summer CO-OP job at General Electric's Gas Turbine engineering group in Schenectady, NY

At this time I used to meditate at my desk during the lunch hour.

One day in early August I was meditating and thinking about a trip I was planning to Cape Cod. My mind was wandering as I was thinking about what I would do there. My thoughts went to what I would do at the beach.

All of a sudden, I had a blinding flash of a scene where I was in the surf at the beach, and a surfboard was coming towards me. Then a shock occurred and I was thrown out of my meditation and was wide-awake.

I thought that this was pretty weird, and mentioned this to a friend or two.

Two weeks later I was walking on the beach on Cape Cod. I saw a couple of guys with surfboards and asked where I could rent one to give it a try.

They said they had an extra one and I could try it with them. (I had totally forgotten my meditation vision at this point)

I tried to get up on that board all day, and had some modest success, but I was also getting exhausted in the process.

I decided to try it again and fell off when a big wave hit me. Next thing I knew I was coming up to the surface and I saw the exact same scene from my meditation.

The board hit me hard in the chin and almost knocked me out. I staggered to the shore and the two guys I was with helped me to the hospital where they put 10 stitches and 2 sutures into my chin.

The question arises—Would I have been able to avoid the accident if I had remembered my vision and not gone surfing?

Later experiences have convinced me that the future is a set of probabilities, and we have free will to decide our actions.

I also had an experience on that trip of being able to partially heal my wounds very quickly through a deep meditation and application of psychic healing techniques. However, I do still have a small scar on my chin from this accident.

b.   Warnings of Danger

- Detroit

In 1980 I was moving from Dekalb, Illinois to Rochester, NY between assignments at General Electric, Inc.

While staying with my cousin outside Detroit, I made arrangements one evening to meet an old RPI friend Steve. We decided to go into downtown Detroit to the newly completed Renaissance Center to eat dinner and look around.

The Renaissance Center was built near the water and surrounded by slums.

After dinner we were walking out through the lower level in an area that was all boarded up with nobody else there.

Suddenly, I had this strong urge to turn around and go to find a restroom. I stopped walking forward because the urge was so strong.

I tried to walk forward again and again a very strong urge came to turn around and go back into the main center where other people were. I remarked to Steve that I couldn't go forward—that something wouldn't let me.

Just then two black guys in trench coats appeared about 30 feet away from behind one of the foundation pillars we were about to walk past. They started walking towards us with smiles pasted on their faces.

My friend Steve took off running back into the main area and after a moment or two I figured I didn't know what these guys were carrying under their coats, so I ran too.

In less than 30 seconds we were back in a populated area with Police present, and the two guys chasing us gave us smiles like "next time we'll get you" and took off going the other way.

I had previously always tried to pray to God for protection, and tried to give a subconscious message to my senses to warn me of danger.

I'm convinced that whatever sense or "angel" warned me that evening, I would have been killed or severely wounded if I had continued walking out of the complex with no warning.

- ## At the Border in El Paso

In December of 1987 my Dad drove out to Houston to accompany me on my move to Los Angeles where we were going to start a business together.

The first night on the road we stayed overnight in El Paso, Texas near the border.

Sunday morning I suggested we stretch our legs by taking a walk down to the Border which appeared to be less than a mile.

As we were walking through a rundown area near the border I had a strong sense that we were in danger. This sense continued for several minutes that somebody wanted to hurt us.

I told my Dad we needed to turn around and he agreed.

We got back to the hotel safely, and even though we didn't see any danger, I'm convinced my extra senses picked up something.

- Planning a Trip to Spain

During early August of 1998, my wife and I decided to send her and our kids to visit her mother in Barcelona, Spain.

I was going to buy a ticket separately, and meet them there during early September.

When I started to call the travel agent to book my ticket I had a terrible feeling of fear about taking the flight.

I tried two other times to book the ticket during the week for a September 2nd departure, and each time I got the same strong feelings of fear and death.

I have always prayed and tried to guard myself mentally to avoid disasters, so finally I took the warning seriously and decided not to go at all.

This was very difficult to do since I really wanted to see my wife and kids, and this meant I would be home alone for a month.

Work wasn't an excuse either, since I wasn't doing any really heavy contract work at the time and could easily have taken the time off.

I called my wife and told her my decision, and she was surprised, but agreed for me to follow my instincts.

On September 2nd the Swissair disaster occurred on a plane leaving Kennedy airport in New York, which crashed in Newfoundland Canada with all lives lost.

I would not have originally been booked on that flight, but could have easily ended up on it since I was due to fly through Kennedy airport, and any delay might have caused me to switch planes.

I will never know for sure, but this was a very strong warning.

I should also mention that for several years before this event I had strong feelings that my I would be killed in the near future. After this happened those feelings ended.

c.   A few Seconds Ahead

Sometimes just having sensitivity about what will happen a few seconds into the future will have a positive effect.

Avoiding a car accident at an intersection is one result.

I believe animals have spiritual abilities too.

Here is an example concerning our last dog Apollo. Some years ago he was watching my wife as she planned to start disconnecting a motor inside our dishwasher.

Apollo started barking madly at us (which he never did) and we suddenly realized that we hadn't turned off the power to the dishwasher.

He seemed to be sensing a future event which made him really worried.

d. Dreams of Indian Ocean Tsunami in 2004

Back around the year 2000 through 2004 I was having a series of dreams which were similar but all slightly different.

I seemed to be in a tropical coastal area and at some type of resort. There were lots of people on the beach and there were different types of resorts in each dream.

I had a lot of fear and then it happened. There would be some type of huge wave which crashed over us, or the tide would go out and a huge wave would come in and I would be covered by the wave.

When that happened I never escaped but seemed to be one of the victims.

I recall this type of dream happening at least five to ten times over that multi year period.
Of course the disaster finally happened—the late December 2004 Tsunami of the Indian Ocean which killed at least one quarter million persons.

Scientists researching this event now think this may have been the worst Tsunami disaster in over 600 years in the Indian Ocean.

I have since had a few dreams about Tsunamis hitting Southern California where I live, but none recently…

2.2 Stories About 9/11

a. Experiences about 9/11/2001

There are many reported experiences of premonitions about the major disaster of terrorists striking the two towers on 9/11/2001. Here are a couple of my own experiences: In 1976 during the summer, I had spent the last couple of years going through a spiritual and psychic development class on the side while going to college. My parents lived in New Jersey and we used to go into Manhattan once in a while. This was while the Port Authority towers were still under construction.

I was very interested in them and decided to go to the site to see how far up I could go in the towers. At the time half the building was finished and the public could go up in the elevators. I was able to get to about the fiftieth floor. I entered the floor and it was empty except for the main columns. Going over to the window I wondered how long these towers would last? I decided to try my intuition to get an answer. So I put my hands on one of the pillars next to the windows, closed my eyes and asked "How long will this building last?" I figured it would be at least one hundred years. Imagine my shock when my intuition said to me "Twenty Five years". This didn't make any sense at the time so I figured I must be wrong.

In retrospect this experience showed me that our prophecy abilities can tell us about events many years in the future.

Then in the year 2000 in September I was in New York City again visiting my sister who lived there. I was married at the time and my ex-wife had some intuitional abilities. We were pushing a stroller with our baby son in it along with my sister who had her baby daughter in her stroller. The location was in the financial district next to the twin towers. Suddenly my ex-wife said "I just got a strong impression that many people are going to die here." This is all the detail I remember. We soon forgot the incident and went on to enjoy our visit with my sister.

Some events are so large that they affect the past and many people even in different times around them.

3.0 Visiting Versailles in the Past

Here is a famous story of two women in 1901 who had a visit to Versailles which shifted to the year 1792

Moberly and Jourdain recounted that they had decided to visit the Palace of Versailles as part of several trips around Paris, detailing how, on 10 August 1901, they travelled by train to Versailles. They remembered not thinking much of the palace after touring it, so they said they decided to walk through the gardens to the *Petit Trianon* but after reaching the *Grand Trianon* found it was closed to the public.

They recollected traveling with a Baedeker guidebook, but said they became lost after missing the turn for the main avenue, *Allée des Deux Trianons* and entered a lane, where they bypassed their destination. Moberly reported that she noticed a woman shaking a white cloth out of a window while Jourdain recalled noticing an old deserted farmhouse, outside of which was an old plough.

At this point they described a feeling of oppression and dreariness coming over them after which men who they thought looked like palace gardeners told them to go straight on. Moberly described the men as "very dignified officials, dressed in long greyish green coats with small three-cornered hats." Jourdain recalled that she noticed a cottage with a woman holding out a jug to a girl in the doorway, describing it as a "*tableau vivant*", a living picture, much like Madame Tussauds waxworks.

Moberly did not observe the cottage, but remembered that she felt the atmosphere change. She wrote: "Everything suddenly looked unnatural, therefore unpleasant; even the trees seemed to become flat and lifeless, like wood worked in tapestry. There were no effects of light and shade, and no wind stirred the trees."

The Comte de Vaudreuil was later suggested as a candidate for the man with the marked face allegedly seen by Moberly and Jourdain.

They reported reaching the edge of a wood, close to the *Temple de l'Amour*, and coming across a man seated beside a garden kiosk, wearing a cloak and large shady hat. According to Moberly, his appearance was "most repulsive... its expression odious. His complexion was dark and rough." Jourdain noted "The man slowly turned his face, which was marked by smallpox; his complexion was very dark. The expression was evil and yet unseeing, and though I did not feel that he was looking particularly at us, I felt a repugnance to going past him. They said that another man whom they described as "tall... with large dark eyes, and crisp curling black hair under a large sombrero hat" came up to them, and showed them the way to the *Petit Trianon*.

Moberly said she noticed a lady sketching on the grass who looked at them after they crossed a bridge to reach the gardens in front of the palace. She later described the lady as wearing a light summer dress and a shady white hat with lots of fair hair. Moberly reported that she thought she was a tourist at first, but the dress appeared to be old-fashioned. Moberly came to believe that the lady was Marie Antoinette. Jourdain, however, did not see the lady.

At their return to the palace, they reported that they were directed round to the entrance and joined a party of other visitors. They said that after they toured the house, they had tea at the *Hotel des Reservoirs* before returning to Jourdain's apartment.

Real Time Travel Stories From a Psychic Engineer

4.0 Rudolph Fentz 1951

The Fentz legend describes how one evening in mid-June 1951, at about 11:15 p.m., passersby at New York City's Times Square noticed a man of about 20 years of age, dressed in the fashion of the late 19th century. No one observed how he had arrived there, and he was disoriented and confused standing in the middle of an intersection. He was hit by a taxi and fatally injured, before people were able to intervene.

The officials at the morgue searched his body and found the following items in his pockets:

- A copper token for a beer worth 5 cents, bearing the name of a saloon, which was unknown, even to older residents of the area;
- A bill for the care of a horse and the washing of a carriage, drawn by a livery stable on Lexington Avenue that was not listed in any address book;
- About 70 dollars in old banknotes;
- Business cards with the name Rudolph Fentz and an address on Fifth Avenue;

- A letter sent to this address, in June 1876 from Philadelphia;
- A medal for coming in 3rd in a three-legged race.

None of these objects showed any signs of aging. Captain Hubert V. Rihm of the Missing Persons Department of NYPD tried using this information to identify the man. He found that the address on Fifth Avenue was part of a business; its current owner did not know Rudolph Fentz. Fentz's name was not listed in the address book, his fingerprints were not recorded anywhere, and no one had reported him missing.

Rihm continued the investigation and finally found a Rudolph Fentz Jr. in a telephone book from 1939. Rihm spoke to the residents of the apartment building at the listed address who remembered Fentz and described him as a man about 60 years who had worked nearby. After his retirement, he moved to an unknown location in 1940. Contacting the bank, Rihm was told that Fentz died five years before, but his widow was still alive but lived in Florida. Rihm contacted her and learned that her husband's father (Rudolph Fentz) had disappeared in 1876, aged 29. He had left the house for an evening walk and never returned. All efforts to locate him were in vain and no trace remained.

Captain Rihm checked the missing person's files on Rudolph Fentz in 1876. The description of his appearance, age, and clothing corresponded precisely to the appearance of the unidentified dead man from Times Square. The case was still marked unsolved. Fearing he would be held mentally incompetent, Rihm never noted the results of his investigation in the official files.

5.0 Paul Dienach

Chronicles from the Future: The amazing story of Paul Amadeus Dienach (As told by the storyteller)

Introductions typically attempt to present the essence of a book, highlighting the most important elements of the story you are about to read. My introduction does not do that. Rather, I will be telling you the story of how this unique text came to be, its journey from the 1920s until today.

This is a book that contains the diary of a man who never intended his words to be revealed to the world. It chronicles an experience that was never shared for fear of ridicule and disbelief. As you work your way through his very personal memoire, the reason for secrecy will soon become clear – the author claimed to have lived in the future and returned back to his original era, 20th century central Europe, to record a detailed account, outlining exactly what happened during his journey.

The real protagonists of this amazing, true story are two persons: Paul Amadeus Dienach, the author, and the man who claimed to have lived in the future; and George Papahatzis, Dienach's student of German language studies to whom he left his notes - the diary you hold in your hands today.

After making the first acquaintances, let's start unravelling their story step-by-step.

Paul Amadeus Dienach was a Swiss-Austrian teacher with fragile health. His father was a German-speaking Swiss and his mother was an Austrian from Salzburg. Dienach travelled to Greece in the Autumn of 1922, after having recovered from a one-year coma caused by a serious illness, hoping that the mild climate would improve his condition.

During his time in Greece, Dienach taught French and German language lessons in order to provide himself with a minimum income. Amongst his students was George Papahatzis, a student that Dienach appreciated more than any of the others. Papahatzis describes his teacher as a "very cautious and very modest man that used to emphasize the details".

Dienach, as we learn from Papahatzis, was born in a suburb of Zurich and lived his adolescence in a village near the large Swiss city. He later followed humanitarian studies with a strong inclination to the history of cultures and classical philology. It is believed that he eventually died from tuberculosis in Athens, Greece, or on his way back to his homeland through Italy, probably during the first quarter of 1924.

Before Paul Dienach died, he entrusted Papahatzis with part of his life and soul– his diary. Without telling Papahatzis what the notes were, he left him with the simple instructions that he should use the documents to improve his German by translating them from German to Greek.

Papahatzis did as he asked. Initially, he believed Dienach had written a novel, but as he progressed with translations, he soon realized the notes were actually his diary… from the future!

At this point we have to clarify something crucial. Dienach is thought to have suffered from Encephalitis lethargica, a strange neurological disease that develops an immune system response to overloaded neurons. The first time Dienach fell into a lethargic sleep it was for 15 minutes. The second time it was for a whole year…

During this year that Dienach was in a coma in a Geneva hospital, he claimed to have entered the body of another person, Andreas Northam, who lived in the year 3906 AD.

Once he recovered from his coma, Dienach didn't talk to anyone about his remarkable experience because he thought he would be considered crazy. However, what he did do was write down the entirety of his memory relating to what he had seen of the future. Towards the end of his life, he even stopped his teaching job in order to have as much time as possible to write everything he could remember.

Dienach describes everything he experienced of the environment and people of the year 3906 AD, according to the mind-set and limited knowledge of a 20th century man. This was not an easy task for Dienach. There were many things he claims not to have understood about what he saw, nor was he familiar with all their terms, technology, or the evolutionary path they had followed.

In his memoires, he claims that the people of the future fully understood his peculiar medical situation, which they called " **conscious slide** ", and they told Dienach as many things as they could in relation to the historical events that took place between the 21st and 39th century. The only thing they didn't tell him was the exact story of the 20th century, in case Dienach's consciousness returned back to his original body and era (as he did) – they believed it would be dangerous to let him know his immediate future and the future of his era in case it disturbed or altered the path of history and his life.

By reading Dienach's unique personal narration page by page, you will be able to decode what he claims to have seen in relation to mankind, our planet, and our evolution. Many may wonder – what happened to the diary in all that time, from the distant year of 1926 until now, almost a century later?

George Papahatzis gradually translating Dienach's notes – with his not so perfect German – over a period of 14 years (1926-1940), mostly in his spare time and summer breaks. World War II and the Greek civil war delayed his efforts of spreading the amazing story that landed on his desk all those years ago.

On the Eve of Christmas in 1944, Papahatzis was staying with friends at a house which was also used and occupied by the Greek Army. When the soldiers caught sight of Dienach's notes, which were of course in German, they confiscated them because they considered them suspicious. They told Papahatzis that they would return them only after they had examined their contents. They never did. But by then, Papahatzis had already finished the translation.

George Papahatzis tried to track down information about Dienach, by visiting Zurich 12 times between 1952 and 1966. He could not find a single trace of him, nor any relatives, neighbors, or friends. Dienach, who is thought to have fought with the Germans during World War I, probably never gave his real name in Greece, a country that had fought against the Germans.

After the end of World War II and the Greek Civil War, Papahatzis gave the translated diary to some of his friends – masons, theosophists, professors of theology and two anti-Nazi Germans– and after that, when everybody realized what they had in their hands, the diary was kept within a close philosophical circle and in the Tectonic Lodge, in which he was a member. The book was taken very seriously by the Masons, who did not want the information spread to a larger circle. They considered the book to be almost holy, containing wisdom about the future of humanity, and better kept only for the few.

Finally, after strong disputes, George Papahatzis decided to publish Dienach's Diary. It was during the period that Greece entered the hardest phase of the seven year dictatorship in 1972. Strong protest from certain church circles – who considered the book heretic – and the fall of the dictatorship a year later, condemned the first edition to oblivion. No one was interested in the future when the present was so intense and violent.

All these factors, along with the difficult language and the rough style of Dienach's notes, which mixed together elements of his past, along with his experience of the future, made the diary even more difficult to understand. Only a few had the time, patience, and knowledge to decode the secret knowledge that lay encoded within almost 1,000 pages.

Another edition followed in 1979 in Greece. However, again the book disappeared and it was hardly mentioned again, apart from the few that knew of its existence. After all the silence, Papahatzis died, and his family did not wish to carry on with his work.

Twenty two years passed before the diary was picked up again by Radamanthis Anastasakis, a high ranking member of the Masonic Lodge in Greece, who decided to publish the book on a small scale, exactly as it was previously written.

That's when I discovered the book for the first time and started to "restore" it, without the sentimentalities that kept Papahatzis from doing something more than an exact translation of the 'holy' scripts of his teacher. Almost a century after the original script was written, this was a task that had to be undertaken so that a 21 st century reader could really understand what a 20 th century man wanted to say.

And so I did it, making sure not to change any of the content, but filtering out irrelevant notes pertaining to Dienach's early life and emphasizing his experience of the future, but in a simpler language and without the gaps that Dienach's narration had.

I have tried to keep the true essence of his story intact. This was my debt to Dienach, whose chronicles of the future completely changed my perspective of life. Nothing more, nothing less. My only goal was to make it accessible to all of you, because if Dienach's experience was indeed real, this book contains revolutionary information – something the Masons clearly recognized – and has the potential to radically change your view of the world and mankind.

Now that you know the background to this unique story, I will simply deposit the future in your hands with an abstract from the introduction of the 1979 edition of the book by George Papahatzis, the man who personally knew Dienach:

The translator of the original texts, knew Dienach personally. His belief is that the inspiration and writing of these texts wasn't an imaginary creation of Dienach, based on his education and insightful abilities. It is a true phenomenon of parapsychology that was linked to his life. Maybe he has also added his own things, maybe he didn't see or live all of the events that he so vividly describes and presents. What is certain is that most of the basic elements of his texts are true experiences that he had; he lived in advance a part of the future to come and a metaphysical phenomenon of incredible clarity happened to him - a phenomenon of parapsychology that rarely happens with such an intensity and roughness. Because of him, what is going to happen on Earth starting from the last decades of the 20 th century up to 3906 AD, is now known to us, at least in general terms.

I have to tell you that while Papahatzis was just a student at the time of receiving Dienach's diary, he went on to become a very respectable man of his era. He was Vice President of the European Movement (National Council of Greece), Founding Member of the Greek Philosophical Society, and a Professor of Philosophy and Culture. He risked a lot in publishing Dienach's work and this on its own reflects his unwavering belief in its authenticity.
Now I leave you with Dienach's diary, a chronicle from the future... 23rd May 2015

6.0 Bold Street, Liverpool, England

Bold Street is a site with the most experiences of time slips so far discovered. The below stories indicate that there is some type of time warp on this street which some people pass through.

The Liverpool Time Slips and Mysterious Occurrences in Bold Street are numerous. This location seems to be some type of time portal. Several stories follow.

The Bold Street Timeslips Liverpool parascience.org.uk

The subject of time has always intrigued us. Is it as set as we have always believed? Or does time loop back on itself, giving us a glimpse of a shadowy past out of the corner of our eye.

Was is just our imagination that made us believe we had seen an object or building change before our very eyes, and seem as though we had stepped back into the past? When this happens we usually shake our heads and put it down to imagination.

But over the last few decades, something strange has been happening in or near Bold Street, Liverpool England. Not just a glimpse of the past, but full immersion into the strange and mysterious world of English History, if only for a few moments at a time.

The strange thing about the Bold street time slips is the actual time and place they are set. In the following cases, the people involved do not go back really far, but seem to visit a particular decade or decades.

So far, most of the sightings have centered around the 1950s and '60s. This is strange in itself. Most time travel experiences seem to take the recipient back to the 18th or 19th century. But not in this case.

Are these people simply copying each other in their experiences, or are they genuinely taking a step back in time?

The answer to this has to take into account whether they are doing it deliberately to get noticed. In other words are the perpetuating a hoax?

Another explanation could be mass hallucination.
And last but not least, they really are experiencing this strange phenomena!

The most important point is, the very first person that had this experience, obviously totally believed in what he saw, heard and felt.

So, does time flow like a river? Or does it twist and turn, going forward then sweeping back, picking up historic events and placing them down in front of you, if only for a few moments?

In this first tale, we find Frank and his wife out for a stroll in Liverpool town center. It is 1996.

His wife decided that she wanted to go and buy a book at Waterstone's the large book store, and they started to walk towards the area of the shop.

As they approached Bold Street, Frank decided to go to another shop first, but bumped into his friend, and stopped to chat in the street. His wife went ahead without him.

A few moments later, Frank said goodbye, visited his shop and turned to go back to meet his wife. After reaching Bold Street, he headed on towards the bookstore. As he approached, he glanced up and was surprised to see the name, Cripps above the door. As he was about to cross over to see what was going on, a van swept past him with the name Cardin's on the side. The van driver honked his old fashioned horn and drove past.

Looking around, Frank suddenly realized that things were not quite what they should be. He looked at the cars driving past and realized that they were all old fashioned vehicles such as people would drive back in the 50's and 60's.

And then he noticed the people. Men were wearing hats and macs, and the women were dressed in head scarves, full skirts and had old fashioned hair styles such as women wore just after the war.

By this time, Frank was beginning to feel slightly freaked out. He carried on crossing the road and headed towards the store.

As he got closer he noticed in the window there were handbags, shoes, and umbrellas. Suddenly he saw a young woman looking up at the shop sign. She looked confused.

She was wearing modern clothes and as she saw him approaching, she smiled at him.

Frank went into the shop, closely followed by the young woman. When they entered he was surprised and pleased to see that it had indeed turned back into a bookshop. The young woman smiled, shook her head and said, 'that was strange, I thought it was a new clothes shop!' then she walked away looking extremely puzzled.

This may sound an unlikely tale, but the odd thing about it is that Frank was, in fact, a former Police officer who was used to dealing in facts, and definitely wasn't the type of person who would believe in the paranormal.

Frank never stopped talking about it. Was this a time slip? Evidently Cripps was a women's shop that sold clothes and other goods decades before!

And Cardin's was also a well-known Liverpool firm that owned vans around the same time.

The second story concerns a young girl by the name of Imogen. She had decided to go into Liverpool to buy her sister Abigail a few things for her new baby. Upon arriving she was happy to see a new MotherCare store that had opened up on the corner of Lord Street and Whitechapel.

She wandered around the store, and picked up a few baby items such as cardigans, baby bibs, and gloves. She was surprised to see how cheap the items were, but thought they were on offer as the store had just opened. Taking them to the counter, she tried to pay with her credit card. The staff member looked at her suspiciously, and went off to get the manager.

When she came back, she looked at the card and told Imogen that they didn't take cards. So, disappointed, Imogen went and put the items back as she hadn't any money with her.

When she got home, she told her mother what had happened. Her mother was surprised and really puzzled. 'That store closed years ago,' she said. 'There is a bank there now, in fact that's where I have my account'. Not believing her, Imogen took her mother back to the same place the next day. Sure enough the store wasn't there. It was a bank, just as her mother had told her.

The third tale is of a young man named Sean, who, while shoplifting in Liverpool back in 2006, ran away from a Security Guard and headed down Hanover Street. Trying to shake off the Guard, Sean, 19, turned into a dead end street called Brookes Alley.

By this time he was out of breath and started to get a tight sensation in his chest. He soon realized that actually it wasn't a problem with him, but the atmosphere around him.

He waited for the Guard to come around the corner after him, but he never appeared. So, thinking he had given him the slip, he sauntered back out and started to walk down Hanover Street again. But he soon realized that something was wrong.

The road looked different, and so did the pavement. He noticed cars driving by that looked very old fashioned, and the road works that he knew were there, were now gone. Soon he saw that the people around him were wearing strange clothes. Crossing over to Bold Street, he noticed that there were traffic lights where they weren't before, and bushes growing around the Lyceum, near a bar that he recognized.

He carried on walking. Soon he began to feel that something was not quite right. Then he began to panic. He realized that somehow he had stepped back in time. And the time slip was not going away.

Then he remember his Cell phone. Taking it out of his pocket, he tried to get a signal, but of course it didn't work. Eventually he began to really panic, but soon spotted a kiosk selling newspapers and headed over.

Leaning over the Stand, he took a look at the front page of the Daily Post. There in bold lettering was the date. 18th May 1967.

He wondered what to do. What happens if he can't get back to his own time? What about family and friends? So, speeding up his pace, he reached H. Samuel the Jewelers, and tried his phone once again. This time it worked. Sighing with relief he looked around and realized that he had returned to the present. But the strange thing was, he could still see, down the end of the road, people still walking around in 1967.

By this time Sean had seen enough, and dived onto a bus to go home. When he was interviewed by the local newspaper later, he stated over four times, the exact account.

Now, you may think that Sean was making the story up to escape from the guard. But the strange tale didn't end there. When the Security Guard was interviewed, he stated that when he ran after Sean, and turned down the dead end alley after him, he said that Sean had completely disappeared!

When the newspaper checked out the facts of Sean's story, they found that everything he said was historically accurate.

These three stories are just the tip of the iceberg. There are many tales from around Liverpool that tell of time slips, ghosts and other strange phenomenon. The stories keep coming thick and fast, and of course the more tales, the more likely people will start to believe that they are all being made up, or as the saying goes, Urban Tales. So, what do you think? Real life time slips, imagination, mass hallucination or purely tales that have started out as fun but have turned into the greatest Urban Legends of all time.

My opinion is that, yes, something did happen.

Probably to the first guy, Frank who was just out shopping with his wife. The others? Maybe it was a case of mistaken roads, taking a wrong turning or just a glitch in the person's memory. By the time they get home they totally believe what happened.

Or is it true? There are so many cases concerning Bold Street, and just about anywhere else in Liverpool, that maybe, just maybe they are all living on top of the biggest time slip phenomena in the World.

Real Time Travel Stories From a Psychic Engineer

7.0 Greek Trireme, Turkey 1984

Incident at Miletus. Story is told in the first person by the man who experienced it:

In 1984 a man was travelling round Turkey with a small group of 15 or so people. We had seen some amazing sites in Eastern Anatolia and along the Lycian Coast and we were making our way back to Istanbul along the Aegean coastline. As you may be aware, that part of Turkey has an amazing number of ancient Greek and Roman sites.

On one particularly hot afternoon, we arrived at the remains of the Greco-Roman city of Miletus. I knew nothing of this place but I was looking forward to a leisurely wandering around the ruins.

Miletus was like every other site we had visited up until that time. We had the whole site to ourselves. Well, that was not quite true, there were also lots of goats and sheep, the occasional tortoise and the incessant buzz of the cicadas. But that was about it.

On arrival, I really had the urge to be alone. As my then girlfriend, Jenny, and the rest of the group headed down to the Roman theatre, we walked off in the other direction. I noticed a pink dome rising out of the trees a short distance away and I went to investigate,

'On getting closer, I realized that it was a somewhat derelict mosque. Islamic architecture has always fascinated me so I decided to check it out. Inside it was very overgrown and I decided, for some odd reason, to climb up on to the roof area around the dome. After a difficult and somewhat dangerous scramble, I found myself on the roof.

The view was splendid. I could see across the site and in the distance I could see Jenny and the others sitting in some shade next to the ancient theatre. There was nobody else around. Even the goats seemed to have left me to my solitude.

Then something really strange happened. The only way I can describe it is that the air around me became electric. I felt a tingling all over me, the view I was looking at seemed to shiver like a TV screen losing its signal. Time seemed to stop. As the air calmed down again. I suddenly found I was looking out over an entirely different scene.

Moments before I had been viewing the swaying cotton fields of the Menderes Rivers round plain but now it was different. The flood plain had turned into a wide estuary filled with water, what had been hills in the distance were now offshore islands. The water lapped in as far as the ruined theatre that I could see further down the valley. I really could not take in fully what I was seeing but it was totally vivid, I can still see it now in my mind's eye. But what happened next snapped me back to 1984. As I looked over at one of the islands -- I saw the prow of a ship appear from behind the island, not a modern craft but a type of ancient Greek galley known as a trireme.

It had a distinctive curved shape to the front. This was the image that was to shake me out of the 'dream' and back into 'reality'. The ancient trireme and the strange vista vanished. The electric sensation ceased, to be replaced by the buzz of the cicadas that greeted me as I returned to the 'present'.

It took me a few seconds to recover, my girlfriend, Jenny, then turned up and took a photograph of her weird boyfriend sitting on the roof of a mosque (A photograph I still possess and have now posted on my Blogsite). When I came down from the heights {both actually and metaphorically) I was still in a strange state of disinclination. A state that would take me many days to fully recover from.

When I returned to the UK, I was keen to know more about the original geography of the area. I was amazed (but out really surprised) to discover that what I had seen was exactly how the coastline looked in ancient times. The Menderes River was a large fjord-like inlet from the sea and the hills present today were then, as I had 'seen' offshore islands. Boats, mostly triremes like the one I had seen, would unload right next to the theatre.

But this event had a strange coda. A few weeks later, the Sunday Times ran an article about that part of Turkey. The reporter, a person whose name escapes me, described how when he was near 'the Mosque" at Miletus he experienced a "curious timeless state'. He said nothing more about this but I wonder what he actually meant. And had he experienced some weird time slip as I had? Somewhere in our loft is the original article from 1984. It so stunned me that I kept it.

8.0 Caught On Video

Caught on Video Stories abound about sightings and encounters with time storms, but in this day and age when there seems to be a surveillance camera on every corner and everyone carries a video camera in the form of a smartphone, it seems that at least one time storm should have been captured on video by now. Well, that just may be the case. In 1996 the news department of a local Florida TV station received an anonymous video taken by a security camera at a small factory in Florida. The video was actually the result of multiple cameras feeding the same deck of monitors. The footage shows a man walking towards the rear gate of the factory. He appears to be looking for something. Suddenly a fuzzy white glow, like an electrical cloud, enters the scene and moves to intercept the man.

Apparently, electromagnetic interference disrupts the video signal for a few seconds. When it clears up, the man has disappeared. The video was subjected to painstaking frame-by-frame analysis by a team of experts, including Ted Williams, a physical scientist, John Carpenter, a psychiatrist, Dr. William Schneid, a criminologist and computer analyst Dan Ahrens. All of these men volunteer their time to investigations supported by the Florida chapter of MUFON, the Mutual UFO Network. Based on the key frame, the team could only conclude that the man simply appeared to have vanished just as the white fog enveloped him. The time of the incident was 11:16 pm. A security guard observing the video monitors had seen it all, and went out to investigate. He was however alarmed when it became obvious that the man, a factory employee, was nowhere to be found. A search was conducted, but the man was found to be nowhere on factory grounds. Maybe he just walked off the job?

They called his home, and his family reported that he was not there and should be at work at the factory. About two hours later, at 1:06 AM, the fuzzy white glow once again made its appearance on the security cameras. This time, sure enough, it appeared to deposit the missing factory worker in the same spot from which he had vanished. The video footage shows the white cloud and then, in a fraction of a second, the man reappears, but now down on his hands and knees, obviously in distress, disoriented and vomiting.

A security guard ran out to help. He asked the worker what had happened, but his mind and memory were a blank. He was sent home in a state bordering on shock. He called in sick the next day, but never returned to his job again. The MUFON team spent hours examining the video, looking at every frame, every fraction of a second, but could find no evidence of fraud. They also subjected the video to a variety of tests to rule out special effects manipulation, but none could be found.

9.0 National Airlines 727

It was 1974 and a National Airlines 727 requested landing instructions from the control tower at the Miami airport.

The captain was surprised when he was instructed by a panicked sounding flight controller to put his plane down on an isolated runway reserved for security risks and special situations, such as when planes are hijacked or in trouble. When the jetliner landed without incident the captain, crew and entire compliment of passengers were surprised when security forces rushed onto the plane and hustled everyone off board, sending them through a special disembarking routine. Emergency vehicles, fire trucks and airport police were on hand outside the aircraft. Everything about the flight had been 100% routine, so captain and crew were mystified about what all the excitement was about.

After everyone had calmed down, the captain finally demanded to know what was going on. The head of airport security told him: 'You guys disappeared into thin air for 10 minutes. You might want to take a look at your clock.' He was referring to the jet's on-board chronometer which showed that it read 9.20am. The captain glanced at his wrist watch – but it matched the clock on his plane's control pane. 'Looks alright to me,' he told the security chief.

Then the chief showed him his watch. It read 9.30 AM. As reported in the Miami Herald, an investigation showed not only the onboard clock of the 727 had lost 10 minutes – but so had the personal timepieces of every passenger on board!

Lost Radar Contact

The panic at the Miami airport started when air traffic controllers were looking right at the radar screen blip of the 727, and saw it vanish from the screen. They immediately assumed that the plane had dropped precipitously to the ground in a crash landing, or perhaps exploded in mid-air.

The radar unit was functioning normally and was still registering other flights in the area. Helicopters and rescue units were sent to a swampy area west of Miami where they were certain the National Airlines plane had gone down. For ten frantic minutes, no one was able to find anything. Then, almost like magic, the jet reappeared on the radar screen and in the exact same location where it had been ten minutes previously. During those ten minutes, the blip should have moved considerably from one position on the screen to another. Furthermore, several other planes passed through the area where the vanished plane had been. From the point of view of pilot, crew and passengers, absolutely nothing had happened.

If they lost ten minutes, it had gone completely unnoticed by anyone. The pilot and co-pilot said they experienced no break in communications with the control tower.

In short, they experienced absolutely nothing abnormal. But the fact remains the National Airlines flight did not exist for a period of ten minutes. It was more than a matter of a radar malfunction – after all, where was the plane? It should have landed ten minutes before it did. If it had stayed in the air and strayed off course, the pilots would certainly have been aware of it. The case of the National Airlines 727 has never been solved. It has been written about in a number of books. Many suggest there is only one logical explanation for the event based on the observed facts: The airplane winked out of existence for ten minutes, did not exist, but then reappeared.

It somehow skipped across ten minutes of time. Did it enter some kind of space-time anomaly in the air? Was there a wrinkle in the fabric of reality? The true answer may never be known.

10.0 The Kersey Time Slip

Was the Kersey experience a slip in time? Or maybe it was a case of 'retro-cognition'? Either way, the bizarre experience of three young British Navy Cadets in 1957 would haunt them for the rest of their lives. Their case would attract the attention of multiple newspapers, and was even written about in the stodgy and highly respected American periodical, Smithsonian Magazine. The case also was prominently featured in a book by Scottish writer Andrew MacKenzie.

The events took place in the Barbergh District of Suffolk in the eastern UK. Three young men, all aged 15, had recently signed up to join the British Royal Navy. As part of their early training, they were set with a map-reading and coordinates-finding task. They were to navigate on foot across about five miles of the English countryside. As it turned out, their target location was the tiny village of Kersey in Suffolk. None of the three boys were familiar with the area. William Laing was from Perthshire, Scotland, Michael Crowley from Worcestershire and Ray Baker was a native of East London.

Nevertheless, they had successfully followed the map coordinates given to find where they were supposed to be – the village of Kersey.

Upon approaching Kersey, an almost overwhelming feeling of strangeness came upon the three adventurers, and it wasn't a good sensation. The boys said an oppressive mood invaded their minds. All three agreed that a penetrating sense of depression and subtle fear had inexplicably descended upon them. They also described it as 'a great sadness'. What's worse, they had the tingling feeling they were being watched by people who were hiding, people who were both fearful and unfriendly. This is odd considering they were approaching a lovely, peaceful and picturesque village known for its delightful Old World charm and welcoming attitude toward visitors. But on this day, the streets of Kersey seemed deserted. There was also no sign or movement of anything modern – no automobiles or even bicycles.

There was no drone of an airplane, no TV antennas on roofs, and they could see no telephone poles, wires, or streetlights. Attempting to marshal their feelings of gloom and dread, the boys sauntered into the village and were struck by its shabby appearance. All of the houses seemed decrepit and crude. In fact, many were in a state of disrepair. One of the youths later described them as looking 'hand built of rough-hewn lumber and timber framed.' The village was medieval in appearance. The strangeness of the village caused the boys to recheck their coordinates. Perhaps they had wandered into some kind of long-abandoned dwelling area? They doubted that they had actually found Kersey, but they concluded that they simply must be in the right place. The boys approached one of the nearest buildings which they said was fitted with smallish grimy windows. They pressed their faces to the glass and saw inside what looked like a filthy butcher's shop. They saw the skinned and butchered carcasses of oxen and found the appearance of the partially carved animals to be revolting.

These carcasses were green with mold. There was no other furniture inside the butcher shop, and no people were evident. It was as if the animals had been partially butchered then abandoned. They also recalled other aspects of the 'butcher shop' structure – it had a crudely painted green door and small windows that were coated with grime and grease.

They could not imagine that health inspection authorities would ever allow conditions such as these in any location processing meat or food for general consumption. They moved on to another house which also had small windows and which was also caked with a film of muck. Inside they saw crudely white-washed walls, and again no furniture. The rooms were cramped and not of modern design. They began to get the feeling they were in some kind of 'ghost town' – and even though they encountered no people, they felt certain they were being watched. They felt that, for some reason, everyone was hiding from them, as if they feared them. By now they were thoroughly spooked. They turned and hurried their way out of the eerie village. They climbed a small hill and did not turn around until they reached the summit which was a considerable distance from the 'haunted village'. But when they surveyed the village from their vantage point on the hill, the whole setting seemed to have changed. They noticed smoke coming from chimneys and they heard the peel of a church bell drifting out from the town. They thought they saw people moving about in the streets. In short, it looked like a normal small English country town. All three had had enough, however.

As a feeling of eerie dread was still upon them, they turned and ran. They said it took several hundred yards for their fear and depression to leave them – and they didn't want those feelings to return – so they kept going and returned to their base of operations. They told their supervising officer what they had experienced. He doubled checked their coordinates and confirmed that they had actually been in Kersey. He also 'laughed off' the boys' tales of a medieval village where a modern town of 1957 should have been.

In the years and decades that followed this 1957 experience, what they had gone through continued to trouble two of the three men, although one of them, Ray Baker, more or less forgot about it. But William Laing and Michael Crowley found they couldn't get away from what had happened to them. As it transpired, in the 1980s they both found themselves living in Australia. They contacted each other and began to exchange letters about what they had experienced that day. For William and Michael, the experience still resonated deeply – they wanted to dig deeper for a possible explanation. To this end, they decided to contact the Society for Psychical Research in London.

WEIRD SUFFOLK: the unnaturally still village frozen in time where something deeply unsettling happened to three young cadets

Church Hill, Kersey in the 1950s.

It's one of Suffolk's creepiest stories: three young Navy cadets chanced upon a village that appeared to be stuck in time, and where they felt they were being watched.

It was Suffolk's own Picnic at Hanging Rock, a creepy village where time stood still and where invisible eyes watched three young men as they battled nausea and fear.

On a bright, clear Autumn morning in October 1957, something deeply unsettling happened in Kersey near Hadleigh. Three 15-year-olds, William Laing from Scotland, Ray Baker from London and Michael Crowley from Worcestershire, were taking part in an orienteering exercise on a Sunday morning, steadily crossing the undulating countryside. The Royal Navy Cadets were looking for a waypoint before heading back to base camp to report to their superiors: and as they were close to their quarry, across the wind they heard church bells. At the top of the hill they were climbing, they saw smoke rising from chimneys and the spire of a nearby church - walking towards the village, they began to feel uneasy: try as they might, they could not hear any noise other than the gentle trickle of a stream.

The birdsong that had accompanied their journey had disappeared and the church bells had fallen silent - and that wasn't the only troubling thing. Autumnal trees had been replaced with vibrant green leaves as if it was springtime, there was no hint of the breeze they'd been walking in and the smoking chimneys and church spire had vanished, replaced by timber-framed buildings. There were no streetlights, no TV aerials, no cars and no telephone wires - and there was no sign of any people, anywhere. The only living creatures the boys could see, other than each other, were the ducks that splashed silently in the stream. Filled with unease, the boys began to look around: there in a butcher's shop window were skinned oxen, green with age and covered with cobwebs as if the butcher had left in a hurry, weeks earlier. Houses in the village were bare of furniture, just empty, cold shells.

Just then, a shiver passed through all three youths as all felt the icy stare of invisible watchers from all around the village tracing their every step. It was the last straw. Petrified and nauseous, they walked quickly up the village street, eventually pelting away from the strange, medieval-looking houses, pausing only to glance back to check if they were being followed: at which point they saw smoke rising from chimneys and the spire of a nearby church... Decades later in 1990, Laing - then living in Australia - flew to England to visit psychical researcher Andrew Mackenzie to investigate what had happened and return to Kersey. There, Mackenzie revealed to Laing that his research had uncovered that the building where the three boys had seen the rotting meat had been registered as a butcher's shop from 1790 to 1905 and could possibly have been associated with the trade for centuries.

Laing recalled what he'd felt back in 1957: "It was a ghost village, so to speak. It was almost as if we had walked back in time... I experienced an overwhelming feeling of sadness and depression in Kersey, but also a feeling of unfriendliness and unseen watchers which sent shivers up one's back... I wondered if we'd knocked at a door to ask a question who might have answered it? It doesn't bear thinking about." Mackenzie was puzzled by the fact the church - which dates back to the 14th century - had not been visible during the possible "time slip" and put forward the suggestion that the three had stumbled into Kersey shortly after the plague had killed half its population.

Others, of course, have suggested that Kersey simply looked old-fashioned to three lads, and that over the years, the 'other-worldliness' became something altogether more sinister in the retelling of the story.

But, as Weird Suffolk knows, the county is no stranger to time slips - we covered a curious case at Rougham where a grand red-brick Georgian mansion has been appearing and disappearing since the 1860s. Could something similar have happened just 16 miles away at Kersey?

11.0 Sir Victor Goddard

A Flight through Time

Sir Victor Goddard's trip into the unexplained involved an airplane flight. This was a much more personally harrowing experience.

In 1935, while a Wing Commander, Goddard flew a Hawker Hart biplane to Edinburgh, Scotland, from his home base in Andover, England, for a weekend visit. On the Sunday before flying back, Goddard visited an abandoned airfield in Drem, near Edinburgh, this location being closer to his final destination than the airport at which he landed. The Drem airfield, constructed during the First World War, was a shambles. The tarmac and four

hangars were in disrepair, barbed wire divided the field into numerous pastures, and cattle grazed everywhere. It was now a farm, and completely useless as an airfield.
On Monday, Goddard began the flight back to his home base. The weather was dark and ominous, with low clouds and heavy rain. Goddard was flying in an open cockpit over mountainous terrain without radio navigational aides or cloud flying instruments. Rain began beating down on his forehead and onto his flying goggles badly which obscured his vision. He thought he could climb above the clouds, but he was wrong. He made it to 8,000 feet, looking for a break in the clouds. There was none.

Suddenly Goddard lost control of his plane. It began to spiral downward. He struggled with the controls. He could speed up or slow down, but he could not stop the spin. He was unsure of his location, but knew he was falling rapidly and might smash into the mountains before coming out of the clouds. The sky became darker, the clouds turning a strange yellowish-brown. The rain came down even more heavily. Goddard's altimeter showed he was only a thousand feet above the ground and dropping rapidly. At two hundred feet and still spiraling downward, he began to see a bit of daylight through the murky gloom, but his spiral toward seemingly inevitable death was far from over. Goddard was now flying at 150 miles per hour. He emerged from the clouds over "rotating water" that he recognized as the Firth of Forth. He was still falling.

Suddenly, he saw directly before him a stone sea wall with a path, a road, and railings on top of it. The road seemed to be slowly rotating from left to right. The cloud cover was down to forty feet. Goddard was now flying below twenty feet and was within an instant of tragedy. A young girl with a baby carriage ran through the pouring rain. She ducked her head just in time to avoid Hart's wingtip. Goddard succeeded in leveling out his plane after that. He barely missed striking the water after clearing the sea wall by a few feet.

He was now flying only several feet above a stony beach. Fog and rain obscured all distant visibility, but Goddard somehow located his position. He identified the road to Edinburgh and soon was able to discern, through the gloom, the black silhouettes of the Drem Airfield hangars ahead of him, the same airfield he had visited the day before. The rain became a deluge, the sky grew even darker, and Goddard's plane was shaken violently by the turbulent weather as it sped toward the Drem hangars-and into a different world.

Suddenly, the sky turned bright with golden sunlight. The rain and the farm had vanished. The hangars and the tarmac appeared to have somehow been rebuilt in a brand-new condition. There were four planes lined at the end of the tarmac. Three were standard Avro 504N trainer biplanes; the fourth was a monoplane of an unknown type-the RAF had no monoplanes in 1935. All four airplanes were bright yellow. No RAF airplanes were painted yellow in 1935. The airplane mechanics were wearing blue overalls. RAF mechanics never wore anything but brown overalls when working in hangars in 1935.

It took Goddard only an instant to fly over the airfield. He was only a few feet above the ground-just high enough to clear the hangars-but apparently none of the mechanics saw him or even heard his plane. As he sped away from the airfield, he was again engulfed by the storm. He forced his plane upward, flying at 17,000 feet and then, for a time, at 21,000 feet. He managed to return to his home base safely.

Goddard felt elated when he landed. He then made the mistake of telling fellow officers about his eerie experience. They looked at him as if he were drunk or crazy. Goddard decided to keep silent about what had happened to him. He did not want a discharge from the RAF on mental grounds.

In 1939, Goddard watched as RAF trainers began to be painted yellow and the mechanics switched to blue coveralls. The RAF introduced a new training monoplane exactly like the one he had seen in his flight over Drem. It was called the Magister. He learned that the airfield at Drem had been refurbished.

Another twenty-seven years went by, but Goddard never forgot what had happened. He played it through over and over in his mind. It was not until 1966 that he wrote of this experience. Over the years he had become convinced that there was no way he could have known that the RAF would change the colors of their trainers and their mechanics' overalls four years before these changes took place. Goddard finally concluded that he must have glimpsed the future-or even traveled into it-for a brief moment in time.

12.0 Mel Riley and the Indians

A spontaneous Time Travel experience.

When Mel Riley was a boy growing up in a poor neighborhood of Racine, Wisconsin, he frequently escaped his dreary urban environment for long forays into the Wisconsin woods. There he could spend hours, sometimes days, just existing free in the wild — sometimes even foraging natural forest foods and building crude shelters to stay warm at night. No department store tents or packed lunches for Mel Riley!

On one such excursion when he was a boy around twelve years of age, Riley was in the vicinity of a ploughed corn field. But this was also an area known to be rich in Native American artifacts. Many arrowhead and other implements had been found there, left behind by the Indians since the Stone Age.

Walking along, Mel suddenly detected the pungent smell of wood smoke and cooking food. He also heard voices. He looked around expecting to see some campers but was stunned to see that the field of ploughed black soil was now a grassy plains area - and it was strewn with Indian lodgings with smoke coming out of them — as if this had been an established settlement for years! The only problem was, none of this had been there just minutes before.

Mel said there were cooking fires near the lodges and native peoples going about their lives, as if this was another era in time - and he concluded that it must have been just that. Mel felt that he had somehow slipped back into time, but the experience lasted only for a few minutes. An instant later, the ploughed field returned and the

Indians were gone.

Mel told freelance science writer Jim Schnabel:

'Things just sort of dissolved and there was the field again. It definitely convinced me that we could access different times, different places. Once something like that opens up to you, you can say, I can travel in time, like a time machine. I can go anywhere I want.'

At the time, however, Mel Riley didn't know how it happened, or how he did it. He was convinced that he had literally accessed another century in time, but that it was a spontaneous event. There was nothing in his current understanding as a boy of the early 1950s to explain something which he was nevertheless convinced had been very real.

13.0 Goethe Meets Himself

Goethe was one of the smartest and most well-known writers of the eighteenth century. He was a well known German writer and statesman. In his autobiography he relates a very unusual story:

As I rode along the footpath to Drusenheim a strange fantasy took hold of me. I saw in my mind's eye my own figure riding towards me, attired in a dress I had never worn, — pike-gray with gold lace.

I shook off this fantasy, but eight years afterwards I found myself on the very road, going to visit Frederika, and that too in the very dress which I had seen myself in, in this phantasm, although my wearing it was quite accidental."

The reader will probably be somewhat skeptical respecting the dress, and will suppose that this prophetic detail was afterwards transferred to the vision by the imagination of later years*

14.0 Wesleyan University Time Shift

The legend of the sighting of Clara Mills started in 1963 on a chilly autumn October morning. It was 8:50 a.m. and Coleen Buterbaugh, a secretary to Dr. Sam Dahl, was searching for a guest lecturer on the building's first floor. Walking down the corridor, Buterbaugh recalls that she heard only the normal sounds of classes changing and of students in the halls.

As she entered the office suite at the north end of the C.C. White Building, she found that all the windows were open and that the room was empty. After taking about four steps into the room however, Buterbaugh had a strange feeling that she was not alone. She smelled a strong odor of musty, stale air, and noticed that it had grown strangely silent in the hall outside.

Then she saw it.

"I looked up and just for what must have been a few seconds, I saw the figure of a woman standing with her back to me at a cabinet in an inner office. She was reaching up into one of the drawers. I felt the presence of a man sitting at the desk to my left, but as I turned around, there was no one there."

Gazing out one of the large windows behind the desk, Buterbaugh noted that the scenery seemed to be that of many years ago. There were no streets, Willard Sorority that stands just across the campus was not there. Nothing outside was modern. Buterbaugh was frightened and left the room.

There is still a controversy as to whom or what Buterbaugh saw that morning in the C.C. White Building. She talked later that day with Dr. Dahl about her experience, and he in turn sought out Dr. E. Glenn Callen, Professor of Political Science.

Without revealing the details of his secretary's experience, Dahl asked Callen if he remembered knowing anyone at NWU who might fit Buterbaugh's description of the vision. She had described the shadowy figure as being a tall, young slender woman with black hair. She wore a long-sleeved white blouse and an ankle-length skirt. She had not seen the woman's entire face.

Callen remembered knowing a Nebraska Wesleyan professor who might fit Dahl's description. Her name was Clara Mills.

Clara Mills taught at Nebraska Wesleyan after receiving a bachelor of music degree from the American Conservatory of Music in Chicago. She taught piano, music theory and history of music for 28 years at Nebraska Wesleyan until April 12, 1940.

On that day, she was found dead in her office, apparently the victim of a heart attack. Mills is buried in Lincoln's Wyuka Cemetery.

Since 1963, para-psychologists, psychiatrists and ghost hunters from all over the country have examined the evidence for the ghost of Clara Mills.

The stories continue today including those of possible Clara sightings in Old Main, the Smith-Curtis Administration Building (former site of the C.C. White Building) and the Rogers Center for Fine Arts.

Real Time Travel Stories From a Psychic Engineer

15.0 The Vision in Fotheringhay church

Jane O'Neill was an English school-teacher, who had been rattled by the sight of a traffic accident, after which she took some time off work to regain her spirits.

It was during this period of convalescence that she and the friend with whom she was staying visited a church in Fotheringhay (or Fotheringay, which appears to be more correct), the village famous for the execution of Mary, Queen of Scots, in 1587 (although I remain unconvinced that this gruesome historical fact absolutely must have something to do with what follows).

Jane took great interest in the church, taking in all the sights. After she and her friend returned to the hotel where they were staying, the friend went on to read (aloud) some more about the church from a guide book.

It was then that this odd story really began. Apparently Jane and her friend, while they had been in the same church, had not seen the same church...

Here's how it all started, in Jane's own words (with a few minor omissions):

In October 1973 I was the first person to arrive at a serious accident; a car had driven head-on into a coach behind which I was travelling. I pulled the passengers from the wreck, waiting with them till the ambulances arrived. Afterwards, with my hands covered with blood, I drove to London Airport to pick up a friend.

Driving home later that night I began to 'see' all over again the dreadful injuries of the passengers. They continued for days. I am usually a very sound sleeper, but I now found I could not sleep at all. The doctor said I was suffering from shock. I was away from school for five days. A fellow teacher invited me for the half-term holiday to her cottage in Norfolk, where several inexplicable things happened. I would be sitting in her living room and would suddenly see very clearly before me a vivid picture. It would last a couple of seconds. (...) I do not remember the sequence of these sights, but I remember them very clearly. After one I told Shirley: "I have just seen you in the galleys." As I hardly knew her I was astonished when she replied. "That's not surprising. My ancestors were Huguenots and were punished by being sent to the galleys".

Two months after the accident Jane and her friend visited Fotheringhay church.

Jane spent some time admiring a picture of the Crucifixion, which was arched at the top, with a dove painted in the middle of the arch. Later, in their hotel, the friend read aloud passages from an essay where it said that infinity is sometimes symbolized by a straight line meeting in arch. Jane remarked that the description sounded similar to the arch on the picture they had seen in the church.

As it turned out, the friend had seen no such picture. Jane decided to phone the vicar to ask him about the picture. She was told - by the postmistress - that the village had no vicar, but the postmistress apparently knew the church as the back of her hand, because she arranged flowers there every Sunday. And the postmistress had seen no such painting either. There was, however, a panel with a painted dove behind the altar.

Two years later Jane and Shirley revisited the church in Fotheringhay. The exterior was as Jane remembered it; but the interior, says she, was such that she had "no recollection at all of having been inside that church. It was much smaller than the one I had been and there was no crucifixion. The dove - to my amazement - was not the one I had seen; this one is in a cloud, and its wings are outstretched, not curved."

Naturally, Jane was much intrigued by these events, and started reading about Fotheringhay church. In Joan Forman's book Haunted East Anglia she discovered that there had been reports of people hearing music coming from inside that same church when it was empty of people.

Eventually, Jane received a written confirmation of what she had seen - or thought she had seen. A historian, Tom Litchfield, described to her in a letter the history of the building. The present church was only a part - a nave - of the original building; the rest had been pulled down in 1533. And a print from 1821, depicting the interior of the church, showed arched panels joined by a painting with a dove with outstretched wings.

In short, what Jane O'Neill had seen seemed to correspond exactly to the appearance of the church as it once was.

16.0 The Disappearing Rougham House

There have been multiple sightings of the Rougham house. Here are some of those stories:

In October 1926 Ruth Wynne was tutoring for a young lady by the name of Evelyn Allington. Ruth's father was the Reverand at Rougham Rectory, and Evelyn's lessons would take place there in the morning and then the girls would take a walk around the surrounding area of an afternoon. On this particular afternoon, they decided to walk to the nearby church of Bradfield St George. Here is their account as told by Ruth Wynne:

One dull, damp afternoon, I think in October '26, we walked off through the fields to look at the church of the neighboring village, Bradfield St. George. In order to reach the church, which we could see plainly ahead of us to the right, we had to pass through a farm-yard, whence we came out on to a road. We had never previously taken this particular walk, nor did we know anything about the topography of the hamlet of Bradfield St. George. Exactly opposite us on the further side of the road and flanking it, we saw a high wall of greenish-yellow bricks. The road ran past us for a few yards, then curved away from us to the left. We walked along the road, following the brick wall round the bend, where we came upon tall, wrought iron gates set in the wall. I think the gates were shut, or one side may have been open. The wall continued on from the gates and disappeared around the curve of the road.

Behind the wall and towering above it was a cluster of tall trees. From the gates, a drive led away among these trees to what was evidently a large house. We could just see a corner of the roof above a stucco front in which I remember noticing some windows of Georgian design. The rest of the house was hidden by trees.

We stood by the gates for a moment, speculating as to who lived in this large house, and I was rather surprised that I had not already heard of the owner amongst the many people who called on my mother since our arrival in the district. This house was one of the nearest large residences to our own, and it seemed odd that the occupants had not called. However, we then turned off the road along a footpath leading away to the right to the church which was perhaps under a hundred yards off. On leaving the church, we cut down through the churchyard into the fields and home, without returning to the road or the farmyard. It was then drizzling rain.

On arriving home we discussed the big house and its possible occupants with my parents, and then thought no more of it.

From this description it's not clear as to which direction they have taken and Carl speculates whether it is in the location of Colesville Grove or the vicinity of Bradfield St. George. Ruth Wynne goes on to say:

My pupil and I did not take the same walk again until the following spring. It was, as far as I can remember, a dull afternoon with good visibility in February or March. We walked up through the farm-yard as before, and out on to the road, where, suddenly, we both stopped dead of one accord and gasped. 'Where's the wall?' We queried simultaneously. It was not there. The road was flanked by nothing but a ditch, and beyond the ditch lay a wilderness of tumbled earth, weeds, mounds all overgrown with the trees we had seen on our first visit.

We followed the road round the bend, but there were no gates, no drive, no corner of a house to be seen. We were both very puzzled. At first, we thought our house and wall had been pulled down since our last visit. But closer inspection showed a pond and other small pools amongst the mounds where the house had been visible. It was obvious that they had been there a long time.

You may think that they had just got lost and had the wrong location, but these women were only walking a distance of about a mile and a half as the crow flies and were doing it on a regular basis. It is unlikely that they were not familiar with their neighborhood.

Map of Kingshall St and possible locations of sightings

James Cobbald

The next account is from James Cobbald (his pen name) he had been told about the disappearing house by another child when he was 11 yrs old. He had laughed at the girl's account and had told his grandmother about it. She then told him that her father; Robert Palfrey, had seen the house. Around 1860 her father had been out in the field making a haystack on a warm June afternoon; as he looked across the field, he could see a house, It was of red brick, and set in a garden with flower beds full of blooms, edged with red bricks placed slantwise. It had two wrought iron gates, one 4 ft wide, the other 9-10 ft. A sudden chill had developed. He had returned home and told his family, and all returned to the location only for the mystery house to no longer be there.

The location was off Kingshall Road and in the direction of
Colesville Grove which is a large grove of dense woodland.
Here is how Carl describes his visit to the grove:

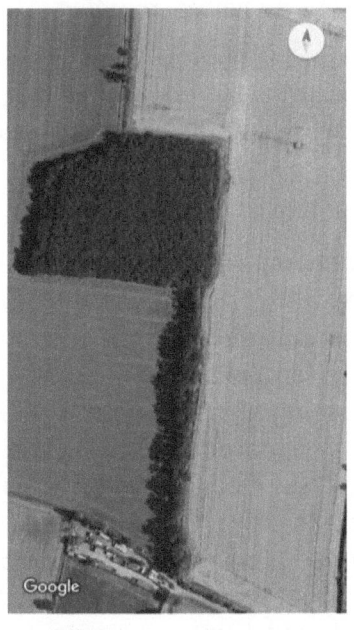

*Most of us have been in woodlands from time to time, but
the Grove is something different. It is overgrown with
every type of nettle, bramble, and thorn. Just moving
around is a major undertaking. In the southern part, where
I entered, the earth forms numerous mounds and ditches,
very like the buried ruins of a large building. Heading north
from the central part is a distinct avenue of trees; the south
avenue contains a few newer trees in what would have
been the driveway. But there can be no doubt about their
significance: nobody would consider planting two*

substantial avenues of trees leading just to a piece of derelict woodland. There must at one time have been a large building at that location. Perhaps not a stately home, but certainly a mansion of some kind. Note that the trees do seem to be around 200 years old, consistent with our deductions from the available mapping.

It was not long after this that James Cobbald had his own experience of the disappearing house. George Waylett, the local pork butcher, was born in nearby Hessett in 1851. He reared the pigs then would slaughter them, and he would bring the carcasses over to his shop in Rougham.

Cobbold would accompany him on his Saturday rounds, making deliveries with his pony and trap. On a warm June day, Cobbold and Waylett were heading south down Kingshall Street when the house suddenly materialized with a loud swooshing noise. The pony uttered a kind of scream of terror and reared up; the butcher fell out of the back of the trap. Then it bolted, and eventually, young Cobbold was able to bring it under control. In those seconds he had had a clear view of a double fronted red brick house. Three storied, of Georgian appearance, and a garden comprising of a large oblong flower bed flanked with two circular beds. And three smaller oblong beds in front, with pansies and geraniums all in bloom, all edged with red bricks placed slantwise, also rose trees. Then a mist enveloped the house, and it faded away. Waylett scrambled to his feet and exclaimed, "That ******** house! That's about the third time I've seen that happen!" Despite Waylett's warnings, the young lad could not resist entering the field and looking in vain for traces left by the mystery building. It is estimated that this sighting took place in 1908

Bentley/Davies sighting

The next sighting was In the 1940s, Edward Bentley was working for the Bury St Edmunds men's outfitter Aubyn Davies. Bentley, aged about 20, used to go out with his manager in the late Summer, distributing catalogues in the surrounding area.

After harvest time, the farm workers had their bonuses and could afford new outfits. Davies was driving, and Bentley and another member of staff were delivering the catalogues. It was a warm, sunny day. They were heading south down Kingshall Street when Bentley suddenly spotted a house off to the right and quickly told Mr Davies that they had missed one. Davies glanced back and reversed the car, but there was now no house to be seen. Bentley put the affair down to a mental aberration, but years later, when discussing the incident with his nephew, Chris Jensen Romer, he realized that he must have seen the ghost house.

He later pointed out the location as the same as the prior sightings by the Grove.

Sandra Hardwick

The next incident was in 1974, Sandra Hardwick was 14 yrs and lived in Rougham, and on a warm summers evening she was meeting her friend at the youth club which was situated at the North end of Kingshall Street (Sandra lived at the South East end of Kingshall) She had promised to get home before dusk and as she was approaching the two bungalows on the east side of Kingshall Street when a house suddenly appeared on her right. It had become unnaturally quiet. The house was brightly illuminated, "like the sun had come out on it on a bright Summer's day." But it was now extremely cold. "I thought I was going bonkers. It was beautiful -- thatched roof, windows open, and a garden with yellow and pink flowers, a fence and a gate." The curtains were blowing out of the open windows. But despite the beauty of the scene, Sandra was terrified, and she pedaled frantically away. Sandra was quoted as saying, "The windows were very small, but open with the curtains blowing, and it was a happy, carefree, friendly house. It had a thatched roof; it was like a perfect country cottage that everyone wants to live in. But there was nobody there."

Jean Batram

On a cool but sunny Sunday afternoon in February 2007, Jean Batram and her husband Sydney (better known as "Johnnie"), a retired couple living in Great Barton, decided to go for a drive around some of the picturesque local villages. They headed south-east towards Rougham, which Jean had never visited before, and drove south down Kingshall Street.

They had just passed the two bungalows opposite Colville's Grove when Jean spotted, on her left side, a large Georgian house. It lay across a newly harrowed field, in front of some woods. She pointed it out to Johnnie, who glanced over briefly, and said that as it was such a lovely house, she would take a closer look at it on the way back.

After a pleasant drive, they returned along the same route. But there was no house to be seen. Jean was puzzled and asked Johnnie if he was certain they had come out on the same road. He told her he was certain as it is the only road running south from Rougham.

Jean became increasingly worried over the coming weeks. She felt that they should report the incident to someone, but Johnnie disagreed vehemently. He declared that he had no wish to be subjected to ridicule, and would deny that he had seen the house himself.

For eight months Jean agonized over the matter. Then, during a phone call to a friend of hers, Katarzyna Powell, she admitted that she had seen something very strange and didn't know what to do. To her surprise, Katarzyna replied, "Oh, you haven't seen the ghost house, have you?" Jean had had no idea that others had also witnessed the same phenomenon. Katarzyna went on to say that her daughter's boyfriend had also seen it, while out driving his van.

The real problem was that while most other witnesses had seen a house on the west side of Kingshall Street, she had seen hers on the east side. Peter and Mary Cornish had told Carl that there was a general disagreement amongst the Rougham community about which side the house appeared on, suggesting, perhaps, that other sightings on the east side had taken place but remained unreported. Mary's grandmother had always told her that the house was seen there. She was certain that it was a fairly large Georgian style house, and that it was standing somewhat to the right of Gypsy Lane, a narrow track which runs from Kingshall Street immediately south of the second bungalow. Carl states "As Phil explained to me, Gypsy Lane is a Greenway, a path originally employed by monks to transport wood to the Abbey at Bury St Edmunds. The Lane is an area subject to unusual events: ghostly figures, strange lighting phenomena, and other interesting occurrences."

85 Kingshall St

View of Gypsy Lane from Kingshall St

Another focus for strange events is Gypsy Lane, the track leading off Kingshall Street immediately south of the two bungalows. The Rose family, who have lived in the second bungalow for many years, have had many occurrences to report. On one occasion, Edith Rose was crocheting in the living room that is right alongside the Lane. Suddenly half of the room became intensely black. When she placed her hand inside the dark zone, she could no longer see it. It is very hard to account for such a phenomenon in any normal physical terms, and I have never come across it before. Could it be a localized time slip back to a night-time period?

Another time, several monks were seen walking past the living room window. Shadowy figures were often seen coming up the front path, always at dinner time. The Roses' horse would refuse to walk past a certain point on Gypsy Lane. Often horses and ponies would break free and run off in a panic. Bob and Win Barker, who lived in a house at the northern end of White Horse Lane, about 400 yards south of the Grove, often independently observed balls of light in their bedroom. The lights would emerge from a wardrobe on the right of the room and travel to the left, about two feet below the ceiling until it disappeared into another bedroom.

An elderly lady and her two daughters, who live in the main part of the village, have a good view of the fields leading up to the Church. On many occasions, they have witnessed a strange arc of light come up out of the ground and form a kind of rainbow. The light persists for a considerable period, but they have been afraid to mention it to others, the last sighting being 2011.

Years ago, when Rougham extended to what is now Moreton Hall, the railway line ran at ground level, and there was a level crossing where today a small bridge exists. One of the callers responding to Carl's appeal in the Free Press, a gentleman named Peter Webb, told him of a rather sinister experience that his father had at that spot.

Cycling over the crossing one evening, Mr Webb had seen what looked like a body lying about 50 yards down the track. It seemed as if a terrible accident had taken place. Dismounting, he walked towards the body, only for it to disappear before he could reach it.

Another story told to Carl by Phil Sage the local historian - he and his wife had just moved into a cottage near the Bennet Arms in Rougham, and he was home on extended leave when one evening a noise on the stairs attracted his attention. His baby daughter was in her cot on the landing, about six steps up. Standing over her was an old lady, wearing a knitted hat. As he watched, with his eyes popping out, the figure faded away towards the window. He was reluctant to worry his wife by telling her about this but mentioned it a few days later to an elderly neighbours three doors away. "Oh, that's nothing to worry about," she declared. "That's just old Millie, looking after your baby. She's often around here." In the context of the Rougham mystery, ghosts are minor players.

17.0 Bampton, England 1993

The village was called Bampton.

The couple drove through it and saw many flowers and a sign saying it had won a competition for 1976. Next day they returned to the same village (or what they thought was the same village) and found it looking different with no flower displays or sign.

That they visited the same village both times. That the sign noting the year 1976 meant it WAS 1976 or soon after.

Bampton might have just had the sign up in 1993 (when the incident occurred). And the village they visited second time around might have been another one (they admit to getting a bit confused).

The most interesting 'clue' in this case was that the couple claim they set fire to their map when getting lost on the first visit. But when using it next day to try to retrace their steps it was not even singed.

This makes you wonder if the 'Timeslips' was a kind of collective vision (a folie a deux) (shared dream/vivid memory) as opposed to a real world visit - thus explaining why the map was never actually burnt.

18.0 The Vanishing Hotel

A widely-publicized case from October 1979, described in the ITV television series Strange But True?, concerned the Simpsons and the Gisbys, two English married couples driving through France en route to a holiday in Spain. They claimed to have stayed overnight at a curiously old-fashioned hotel and decided to break their return journey at the same hotel but were unable to find it. Photographs taken during their stay were missing, even from the negative strips when the pictures were developed.

A full story of this unusual story follows:

In October of 1979, two married couples were driving through France to vacation in Spain. Geoff and Pauline Simpson and Len and Cynthia Gisby decided to find a place to stay for the night near Montelimar, France. The

Ibis Motel was full, and they were told that they might be able to find a room down the road.

The road they followed was narrow and made of cobblestones. They passed buildings, and signs advertising a circus, all of which looked oddly old-fashioned. Soon they came to a long stone building of two storys facing the road. Upon entering the building, the couples found themselves in a large room that appeared to be a bar. The proprietor did not speak English, but was able to communicate that there were rooms available.

The Simpsons and the Gisbys found their rooms to be clean, although very outmoded. The furniture was made of heavy wood, the windows were merely shutters, there were only wooden catches for locks on the doors, the bed sheets were made of calico, there were bolsters with no pillows, the plumbing was antique, and there were no telephones or elevators in the building.

After an evening meal of steak, eggs, fries, and beer, they went to bed satiated and thankful to have found a place to sleep for the night.

At breakfast in the dining room the next morning, three very odd individuals entered the hotel: a woman wearing a long dress and button-boots and carrying a little dog; and, two french policemen in uniforms later discovered to be the styles worn before 1905! Additionally, when Geoff and Len asked the officers about the best autoroute to take, the policemen didn't appear to know the word "autoroute" and directed them to an old road miles out of the way.

When paying the bill upon leaving, it was about 1/13th as much as the going hotel rate at that time. Len believed there must be a mistake, but the manager would take no more cash.

Two weeks later, on their return trip from vacationing in Spain, the couples hoped to stay in the same hotel near Montelimar. They turned off the road at the same point, saw the circus signs that they had seen before, but the hotel was no longer there.

After returning home and developing their vacation rolls of film, they discovered that the pictures they had taken at the hotel were missing...from the middle of the roll! The serial numbers of the frames were consecutive, and there were no blanks in the film.

In 1983, the couples returned to the Montelimar area to search for the hotel. A similar place was found, but the Gisbys and the Simpsons were convinced that it was not the one at which they had stayed overnight in 1979.

Geoff Simpson allowed himself to be hypnotized in 1985 by a psychiatrist to see if there were any other memories in his subconscious about the event, but was unable to add anything new to the experience.

There are a lot of unanswered questions regarding this occurrence. British writer and investigator, Jenny Randles, checked into this event and wonders, if this was some kind of time-slip as some believe, why did the manager accept their modern cash as payment? Why did no one at the hotel seem surprised by their clothing or vehicle? The Simpsons and Gisbys had no answers to those questions. They "only know what happened."

Real Time Travel Stories From a Psychic Engineer

19.0 Battle of Nechtansmere

It was January, 2 1950, and a cocktail party held 10 miles away in the little town of Brechin. This party was attended by Miss E.F. Smith, a lady then aged about 55 who was resident in the village of Letham, under Dunnichen Hill.

According to her own account, Miss Smith left the party late, having consumed an unspecified quantity of those delicious cocktails. Driving conditions were extremely poor. It was pitch dark, and 'a fall of snow had been followed by rain.' Two miles outside Brechin, Miss Smith skidded her car into a ditch. There was, she insisted, no question of [her] skid having been due to her fainting, or other lapse of consciousness, nor [had she been] injured in any way, or concussed. She had to abandon her car, however, and continue her journey on foot – a distance of about eight miles.

The paranormal experience began when Miss Smith was about half a mile from the first houses of Letham village and it continued until she reached them. The time was getting on for 2 AM. Peering ahead, she saw a groups of lights moving in the distance which, as she walked on, gradually resolved themselves into a shadowy group of figures carrying flaming torches.

Miss Smith, 'they were obviously looking for their own dead… the one I was watching, the nearest one, would bend down and turn a body over, and, if he didn't like the look of it, he just turned it back on its face and went on to the next one… There were several of them…. I supposed they were going to bury them. Miss Smith had long come to the conclusion that she had somehow witnessed groups of Pictish warriors of the late seventh century, 685 AD.

20.0 Vanishing Oklahoma White Building

A white Ford pickup pulled up to cattle pasture near Ponca City, Oklahoma, in early Fall 1971, and stopped at a gate. Karl, Mark, and Gordon worked for cattle feed distributor and were sent to this remote area to pick up a feeder.

What they found there has kept them silent for 41 years. "We opened the gate, which was barbed wire with no lock, and entered," Karl said. "We went on the property, which was covered with grass up to and over the hood of the truck." They drove through the tall grass to the tank that sat close to a red barn and got out of the truck.

"We realized the tank was almost half full and too heavy to load," Karl said. We decided to leave and drove around the red barn and we saw a large, two story white house, with no lights in front of us." The trio drove back to the cattle feed company and the boss said he'd drain the tank and they could pick it up tomorrow.

"We went to the location to retrieve the tank the next night," Karl said. "This time we decided to go through the old white big house on the hill and brought our shotguns." They drove onto the property over the path they'd made through the grass the day before and loaded the tank. Then they pulled around the barn toward the house.

What they saw burned into their memories. "It was no longer there," Karl said. "We walked up the hill where it stood and there were no signs of demolition, no foundation, nothing at all. What we all seemed to witness the night before was no longer there. We have talked to each other over the years but none of us can begin to explain this vision."

Did the house teleport away, or were they looking at a house from another dimension or period of time? Further analysis of this case is needed.

21.0 Theories of Time Travel

Lots of interesting time travel and time slip stories are reviewed in this book. There seem to be two types of stories:

The first type of story is when the person involved has a vision of an event or location and they don't interact with anybody in that time.

The second and much rarer story is when persons travel to another time. They are actually there and might interact with the people at that time.

As a result of my own premonitional experiences I've learned a few things. I've learned that the future has a certain momentum which can be altered but it takes energy.

My example for this was when I managed to avoid flying on an airplane which later crashed and killed everyone on board. It was difficult to decide not to go because I had emotional momentum and really wanted to go. From my previous experiences I knew that I should not. When I finally made the decision I felt a lot of relief. My strong feelings that I would die soon—which I had for a couple of years—totally went away.

So the future can be modified but it's not easy. Some major events like the 9/11 terrorist attacks would be difficult if not impossible to alter because thousands of people were involved in causing this potential tragedy to become reality.

So the future has multiple paths even though some of it has such a massive effect on consciousness that modifying those events will never happen.

Having had premonitional experiences myself I know it is really possible for our consciousness to travel to different times. My belief is that we have a core eternal spirit which lives outside of time and space. Therefore, when we get into the right mental state our spirit can perceive different time periods. Our spirit then communicates with our conscious mind to present us with these events. This would explain a lot of experiences of observing different times remotely.

The rarer experience is where the person seems to actually have been transferred to that different era or time.

Some descriptions include observations of strange lights or clouds which appeared before the event. This indicates some type of time warp or actual change of the environment was the cause for this event occurring.

The other possibility is what I would call "temporal teleportation" where individuals actually travel to the site and time in question. These persons actually visited the past. Having written a number of books on the paranormal I know about teleportation experiences. The idea of teleportations happening through time is also possible.

My research shows that these are real experiences in this book and research is needed to provide more data for some better theories to be developed.

The past is also fixed. There does not seem to be any indication of multiple pasts. That we can visit those fixed past times in some circumstances.

I've read a number of time travel science fiction novels which all have different ideas about the way it might work, and which present many conflicting paradigms about resolving temporal conflicts. This is a wonderful subject and I think it is a more complex set of phenomenon than is currently understood.

22.0 Summary

Before I began research for this book I had only seen a few of the stories included here. I was very surprised to find so many stories which were internally consistent and had facts which could be checked about times hundreds of years ago.

These stories also show me that the world is much stranger than most of us can imagine. My experiences with the paranormal and my research on the subject seemed to have been comprehensive when I wrote my paranormal books. Now I see that time travel stories present an even stranger additional phenomenon which leads me to again expand my views of the reality of the world we live in.

What an amazing idea it is that in the right conditions we can actually visit our past and possibly interact with the people in those eras.

It is also important to note that many of these stories came from the United Kingdom. This means there must be many more stories from around the world which we don't know anything about.

I understand that the future can be modified since there are many possible paths. These stories say that the past seems to be fixed and can be visited also.

What a wonderful world of exciting surprises that we live in. Enjoy it.

Martin K. Ettington

May 2020

23.0 Bibliography

1. Moberly–Jourdain incident.
https://en.wikipedia.org/wiki/Moberly%E2%80%93Jourdain_incident. [Online]

2. Rudolph Fentz.
https://en.wikipedia.org/wiki/Rudolph_Fentz. [Online]

3. Bullivant, Richard. *Time Travel Stories.* s.l. : Amazon KDP.

4. Bold Street, Liverpool, England.
https://medium.com/@NellRose1/the-liverpool-time-slips-and-mysterious-occurences-in-bold-street-7a42898c124b. [Online]

5. Ancient Origins-Chronicles From the Future.
https://www.ancient-origins.net/unexplained-phenomena/chronicles-future-amazing-story-paul-amadeus-deinach-003121. [Online]

6. The Kersey Timeslip.
https://www.eadt.co.uk/news/weird-suffolk-kersey-mystery-time-slip-1-6422700. [Online]

7. The Kersey Boys. *East Anglican Times.* [Online]

8. Can You Stand the Truth.
https://canyoustandthetruth.com/sir-victor-goddard-flight-time/. [Online]

9. WEsleyan University Time Shift.
http://hauntsofamerica.blogspot.com/2007/11/haunting-of-wesleyan-university-of.html. [Online]

10. Fotheringhay Church. *http://time-slips.blogspot.com/2011/04/vision-in-fotheringhay-church.html.* [Online]

11. 5 Bizzare Time Slip Cases. *https://www.techeblog.com/5-bizarre-time-slip-cases-where-people-mightve-actually-time-traveled/.* [Online]

24.0 Video Links of Interest

https://www.youtube.com/results?search_query=5+time+sli p+cases

https://en.wikipedia.org/wiki/Time_travel_claims_and_urba n_legends

https://www.youtube.com/watch?v=Gik1Gi5a0os

https://www.youtube.com/watch?v=iFP1G9HUZ0w

https://www.youtube.com/watch?v=702LVl2oY0w

https://www.youtube.com/watch?v=6wamx_aTGZo